What does this painting have to do with math?

American realist Edward Hopper painted ordinary people and places in ways that made viewers examine them more deeply. In this painting, we are in a restaurant, where a cashier and server are busily at work. What can you count here? If the server gave two of the yellow fruits to the guests at the table, how many would be left in the row? We will learn all about addition and subtraction within 10s in *Units of Ten*.

On the cover

Tables for Ladies, 1930
Edward Hopper, American, 1882–1967
Oil on canvas
The Metropolitan Museum of Art, New York, NY, USA

EUREKA MATH²

Great Minds® is the creator of *Eureka Math*®,
Wit & Wisdom®, *Alexandria Plan*™, and *PhD Science*®.

Published by Great Minds PBC.
greatminds.org

Printed in the USA

1 2 3 4 5 6 7 8 9 10   QDG   26 25 24 23 22

ISBN 978-1-64497-086-7

EUREKA MATH²™

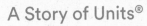

A Story of Units®

Units of Ten ▸ 1

LEARN

Contents

Part 1: Attributes of Shapes

Part 2: Advancing Place Value, Addition, and Subtraction

5 Sides	6 Sides
Pentagon	**Hexagon**

Name

1. Write the number of sides.

2. Trace the shapes.

3. Draw a shape.

Sides	Name	Shapes
	Triangles	
	Quadrilaterals	
	Pentagons	
	Hexagons	

Name

Circle all the shapes with **4** sides.

  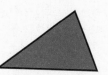 

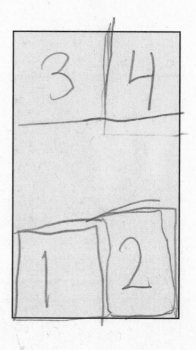

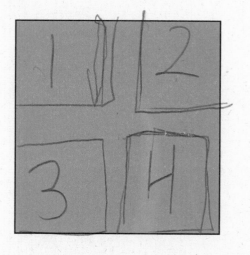

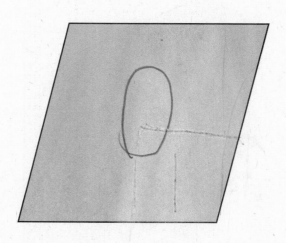

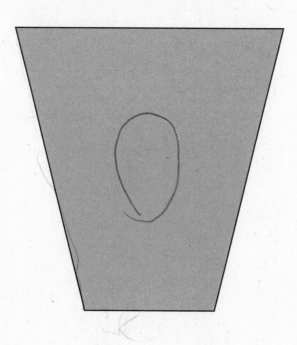

EUREKA MATH²

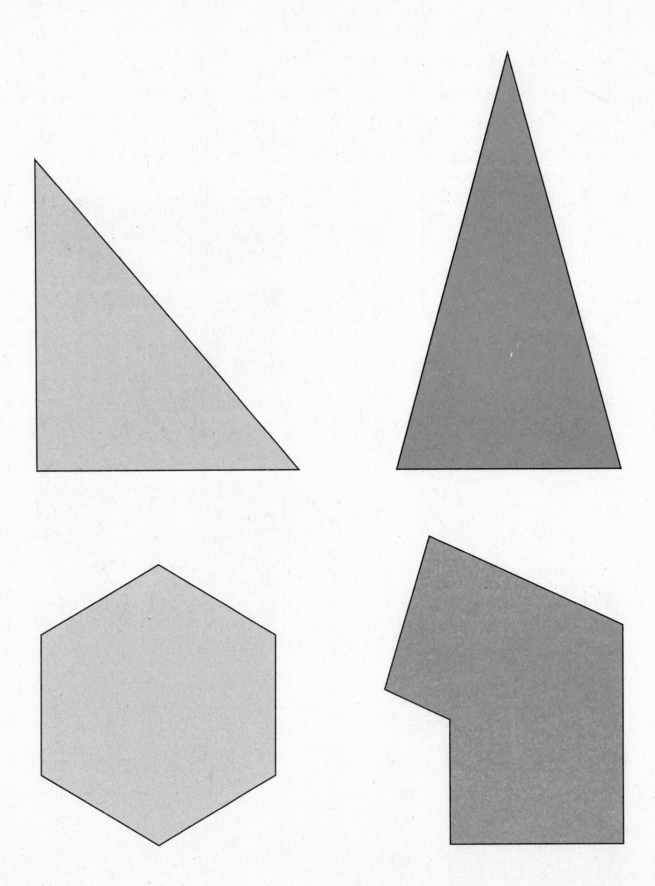

Name _____

1. Count.

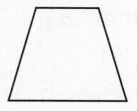

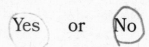

Does it have square corners?

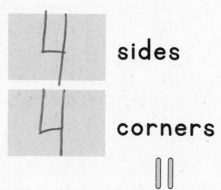

sides

corners

Does it have parallel sides?

Yes   or   No

Yes   or   No

Count.

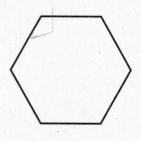

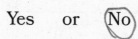

Does it have square corners?

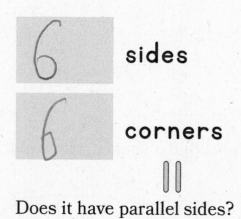

sides

corners

Does it have parallel sides?

Yes   or   No

Yes   or   No

Count.

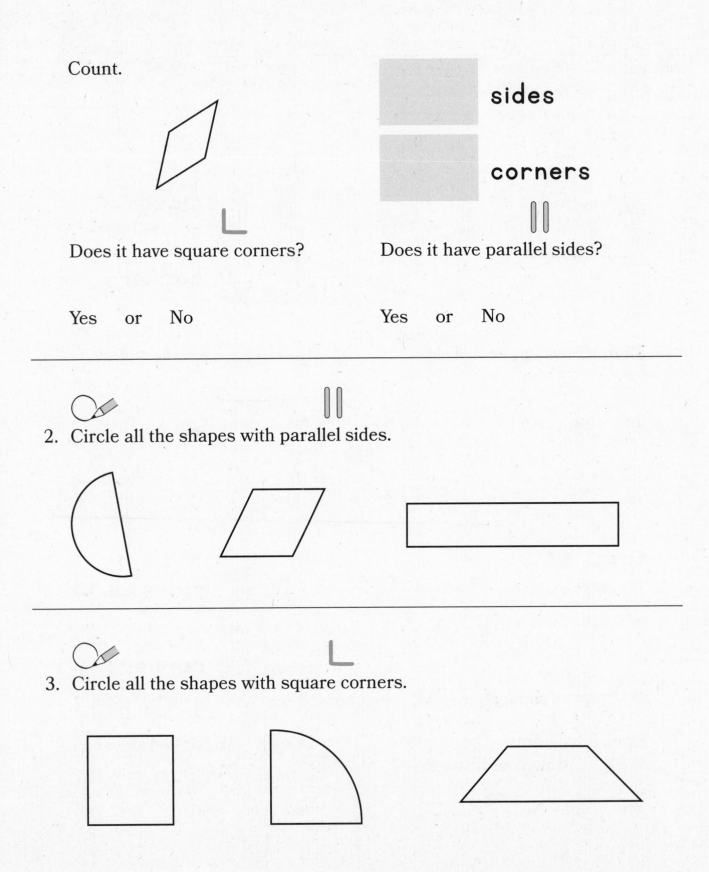

____ sides

____ corners

Does it have square corners?

Does it have parallel sides?

Yes　or　No

Yes　or　No

2. Circle all the shapes with parallel sides.

3. Circle all the shapes with square corners.

PROBLEM SET

4. Draw a shape with parallel sides. Draw a shape with a square corner.

5. Match the shape with its name.

rectangle

trapezoid

rhombus

square

What is the same about all these shapes?

Write or draw your answer.

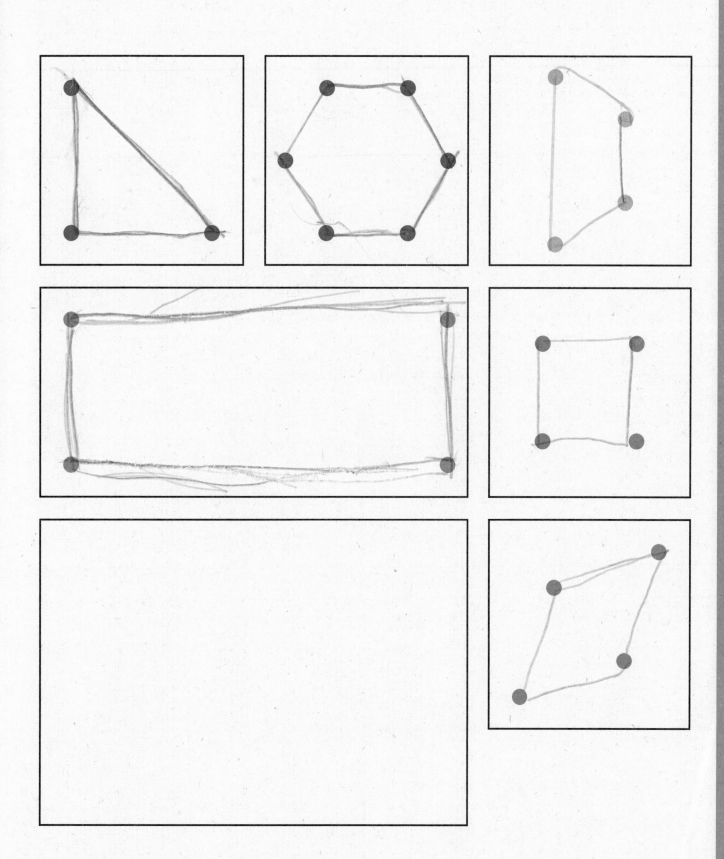

Name _____

Draw a trapezoid.

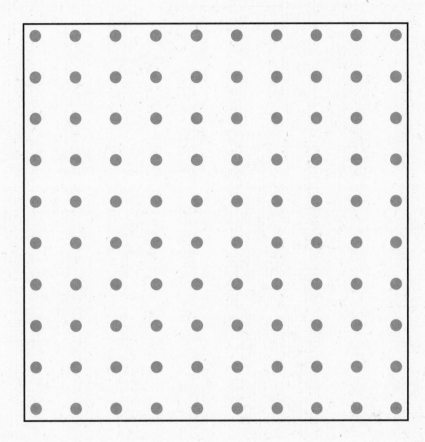

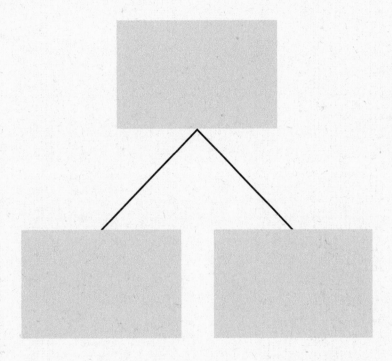

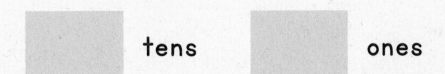

tens   ones

What shape will it be? Circle.

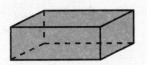 

rectangular prism · cone · triangular prism · cube · cylinder · pyramid

What shape is it? Circle.

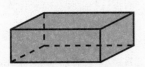 

rectangular prism · cone · triangular prism · cube · cylinder · pyramid

Circle the face shapes. Write the number of faces.

circle · square · triangle · rectangle

faces

Name

Solid Shapes

	cube
	rectangular prism
	cylinder
	cone
	triangular prism
	pyramid

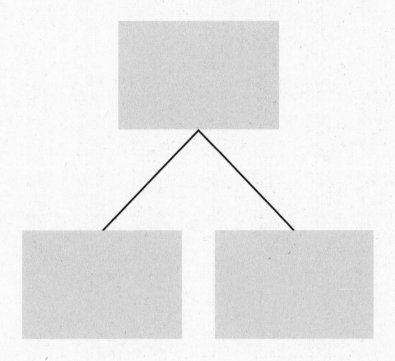

tens   ones

5

Name

1. Circle the shapes that roll.

   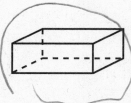  

2. Circle the shapes that stack.

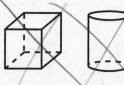 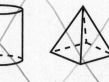 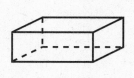   

3. Circle the shapes that do **not** roll.

 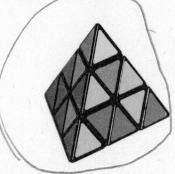 

4. Circle the shape that rolls but does **not** stack.

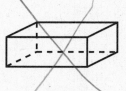   

Name

Sprint

Write the number of tens and ones.

1.	12	1 ten 2 ones
2.	40	4 tens 0 ones

A

Number Correct: _____

Write the number of tens and ones.

1.	11	1 ten 1 one	
2.	13	1 ten 3 ones	
3.	15	1 ten 5 ones	
4.	17	1 ten 7 ones	
5.	19	1 ten 9 ones	
6.	10	1 ten 0 ones	
7.	30	3 tens 0 ones	
8.	50	5 tens 0 ones	
9.	70	7 tens 0 ones	
10.	90	9 tens 0 ones	

11.	31	3 tens 1 one
12.	53	5 tens 3 ones
13.	75	7 tens 5 ones
14.	97	9 tens 7 ones
15.	99	9 tens 9 ones
16.	100	10 tens 0 ones
17.	103	10 tens 3 ones
18.	107	10 tens 7 ones
19.	110	10 tens ___ ones
20.	117	___ tens ___ ones

Name

Squares	Shape	Sides
2	rectangle	4
3	hexagon	6
4	square	4
5	hexagon	6
6	rectangle	4
7	hexagon	6
8	rectangle	4
9	hexagon	6
10	rectangle	

Name

1. Color a rhombus. | Color a rhombus. | Color a rhombus.

What is the last new shape? _____ Sides

2. Color a triangle. | Color a triangle. | Color a triangle.

What is the last new shape? _____ Sides

3. Color 4 triangles.

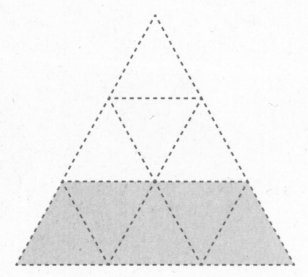

What is the new shape? —————————————— Sides

4. Color in the pattern. How many?

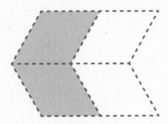

rhombuses

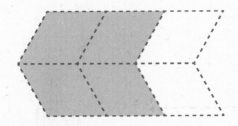

rhombuses

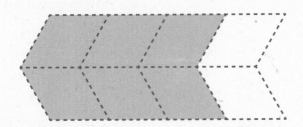

rhombuses

Name

Draw a line to show 2 smaller triangles.

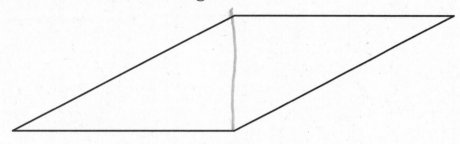

quadrilateral

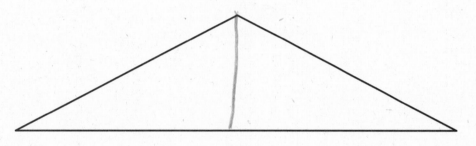

triangle

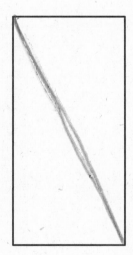

rectangle

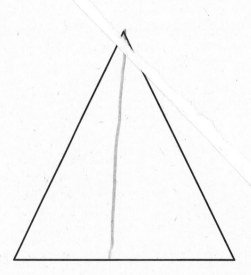

triangle

Name

1. Color 1 square. | Color 2 squares. | Color 4 squares.

What is the last composed shape? _____

It has sides.

2. Color 1 rectangle.

Color 2 rectangles.

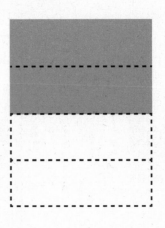

What is the last composed shape? _____

It has  sides.

3. Draw a triangle to compose a square.

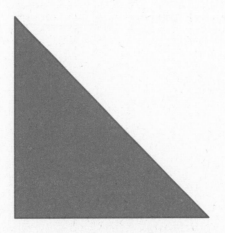

4. Draw a triangle to compose a rhombus.

5. Draw a triangle to compose a larger triangle.

Name

Color 1 triangle.

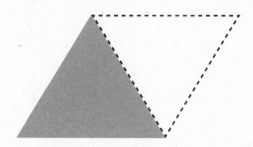

What is the composed shape?

_____

Color 1 triangle.

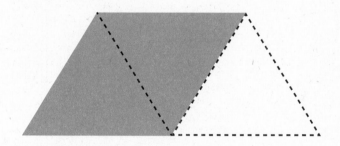

What is the composed shape?

_____

Color 1 triangle.

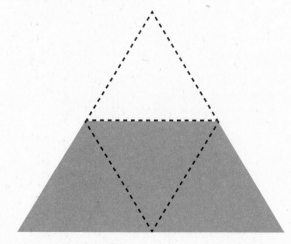

What is the composed shape?

_____

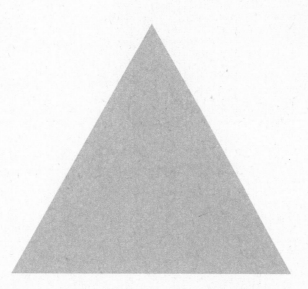

Draw lines to show the blocks you used.

Make the gray hexagon with blocks.

Draw lines to show the blocks you used.

Name _____

1. Circle the triangle made with **fewer** blocks.

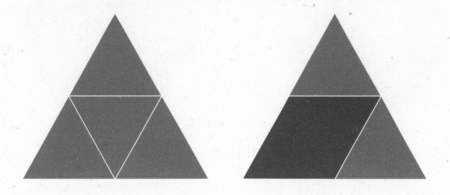

Make a triangle that has fewer blocks.

Draw the shapes.

2. Circle the hexagon made with more blocks.

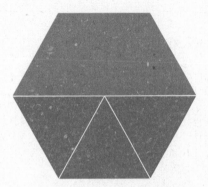 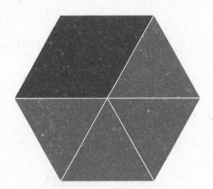

Make a hexagon that has more blocks.

Draw the shapes.

3. Make a trapezoid with blocks.

Draw the shapes.

How many blocks did you use?

4. Make a trapezoid with some new blocks.

Draw the shapes.

How many blocks did you use?

5. Circle the trapezoid that has **more** blocks.

Write why it has more blocks.

_____

_____

EUREKA MATH²

Dom

Name

1 ▸ M6 ▸ TC ▸ Lesson 10

10

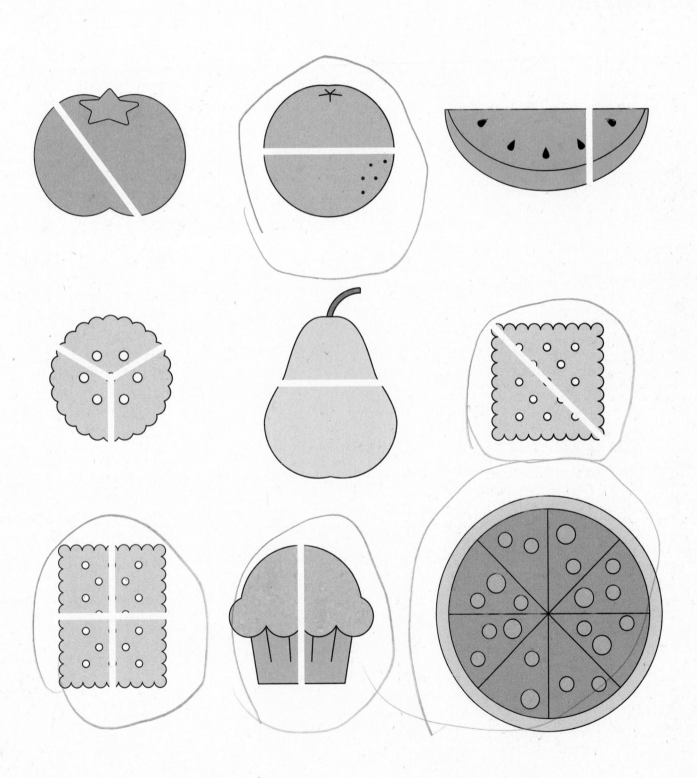

Copyright © Great Minds PBC

95

Name

1. Circle the shapes that show equal parts.

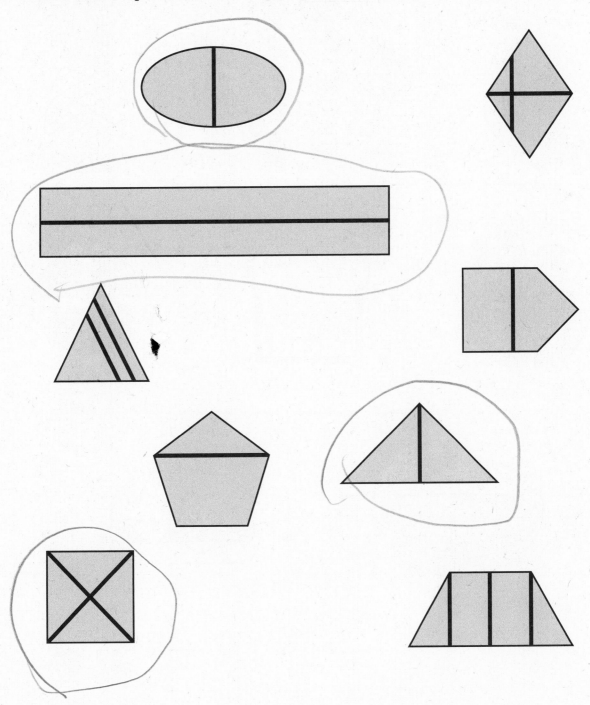

2. Circle the foods that show equal shares.

3. Draw lines to make equal parts.

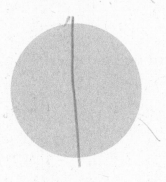

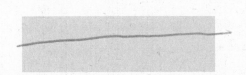

4. Draw a shape.

Draw lines to make equal parts.

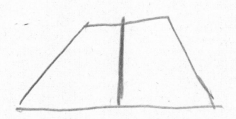

Name Dom

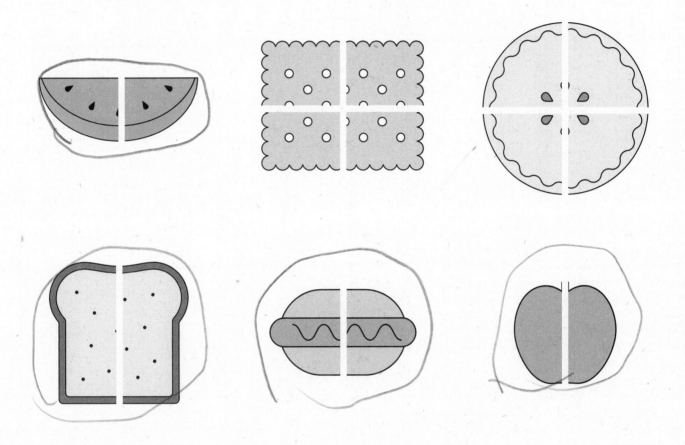

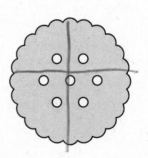

Name

1. Circle the shapes that show **halves**.

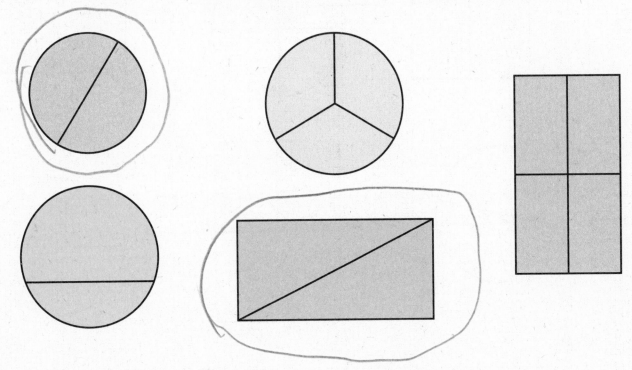

2. Draw to make **halves**.

Color 1 **half**.

3. Circle the shapes that show **fourths**.

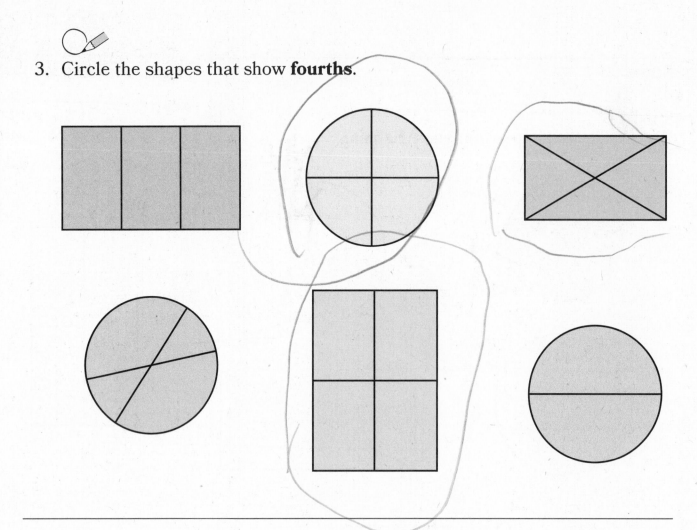

4. Draw to make **fourths**.

Color 1 **fourth**.

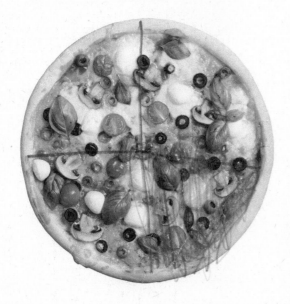

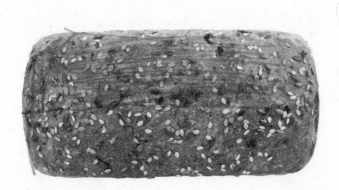

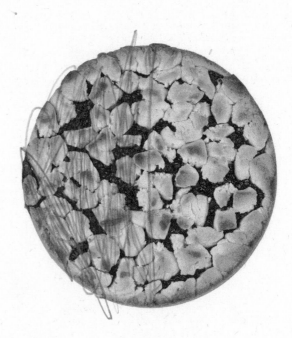

Name _____

Name _____

1. Circle the shapes that show **halves**.

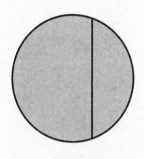

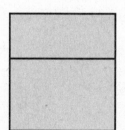

2. Circle the shapes that show **fourths** or **quarters**.

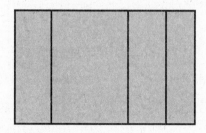

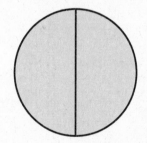

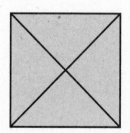

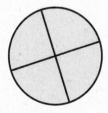

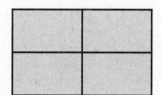

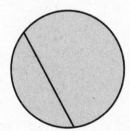

3. Circle.

halves

fourths

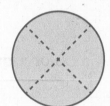   

How many equal parts?

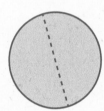

4. Circle.

halves

quarters

How many equal parts?

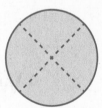

5. Circle.

halves

quarters

How many equal parts?

6. Draw a line to make **halves**.

Color 1 **half**.

Draw lines to make **quarters**.

Color 1 **quarter**.

7. Kit cut a cake. Wes cut a cake.

Who made equal parts? Circle.

Kit   Wes

Write or draw how you know.

Sprint

Write the part or total.

1.	2 + 7 = ▨	
2.	6 + 8 = ▨	

Number Correct: _____

Write the part or total.

1.	$3 + 7 = \blacksquare$	10
2.	$4 + 7 = \blacksquare$	11
3.	$5 + 7 = \blacksquare$	12
4.	$8 + 7 = \blacksquare$	15
5.	$9 + 7 = \blacksquare$	16
6.	$3 + 8 = \blacksquare$	
7.	$4 + 8 = \blacksquare$	12
8.	$5 + 8 = \blacksquare$	13
9.	$8 + 8 = \blacksquare$	16
10.	$9 + 8 = \blacksquare$	17

11.	$3 + 9 = \blacksquare$	12
12.	$4 + 9 = \blacksquare$	13
13.	$5 + 9 = \blacksquare$	14
14.	$8 + 9 = \blacksquare$	17
15.	$9 + 9 = \blacksquare$	19
16.	$1 + \blacksquare = 9$	
17.	$2 + \blacksquare = 11$	
18.	$0 + \blacksquare = 9$	
19.	$\blacksquare + 9 = 15$	
20.	$16 = 7 + \blacksquare$	

Name

1. Color 1 share.

Circle the shape with the **smaller** shares.

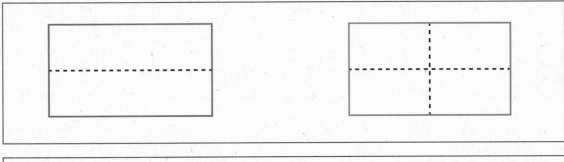

2. Color 1 share.

Circle the shape with the **larger** shares.

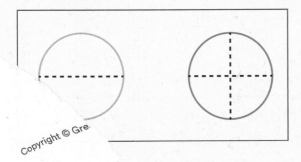

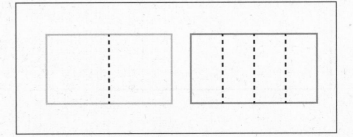

3. Sam cuts 2 equal parts.

Mel cuts 4 equal parts.

 Draw.

 Draw.

Sam eats 1 part.

Color his part.

Who ate the bigger part?

Write or draw.

Mel eats 1 part.

Color her part.

EUREKA MATH²

4:30

1:30

8:30

half past 4

half past 1

half past 8

11:00

5:00

11 o'clock

5 o'clock

Name

Write the times your matches show.

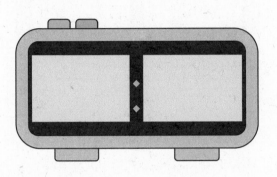 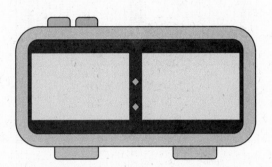

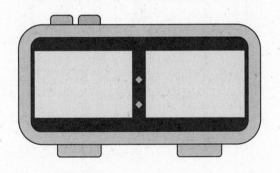 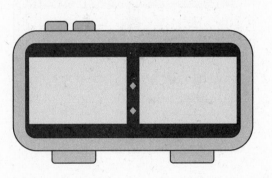

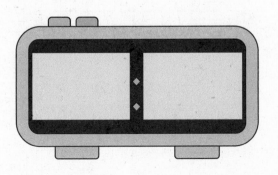 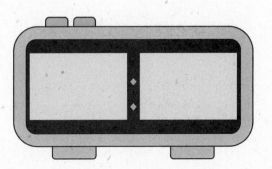

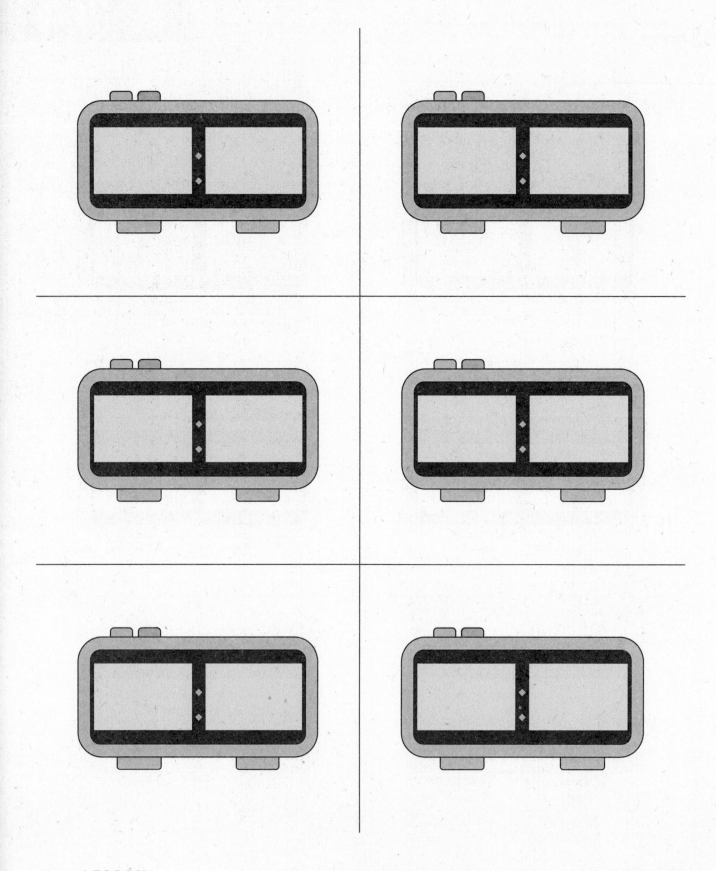

Name

1. Circle all clocks that show half past 3.

2. Circle all clocks that show half past 12.

3. Fill in the blanks.

7:30	half past
2:00	o'clock
11:30	half past
9:00	o'clock

4. Draw lines to match the times.

6 o'clock

6:30

half past 12

half past 6

half past 9

12 o'clock

12:00

5. Draw or write half past an hour on the clocks.

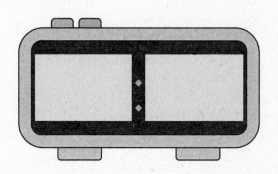

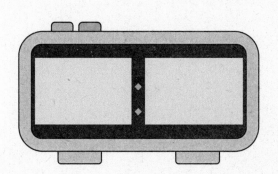

Name

1. Circle the times that match the clock.

7:30   8:30   7 o'clock   half past 7

---

2. Draw to show **halves**.

Color 1 **half**.

3. Draw to show **fourths**.

Color 1 **fourth**.

Name

Write the times your matches show.

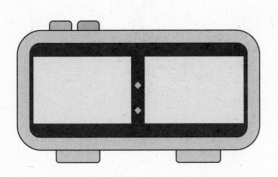

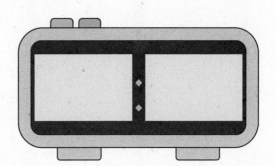

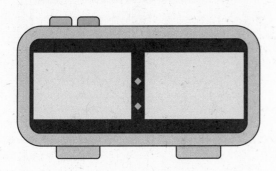

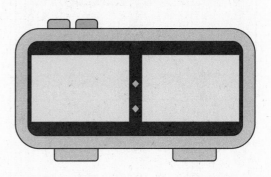

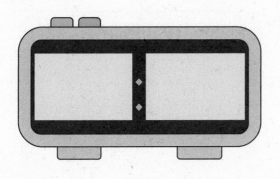

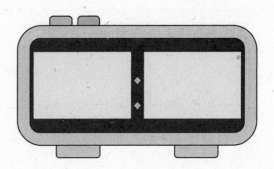

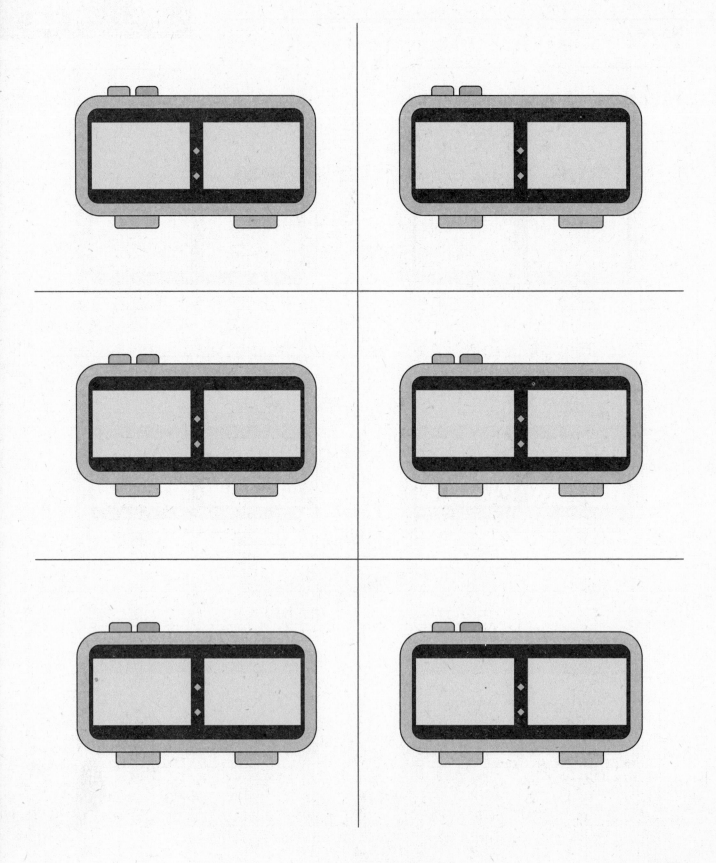

Name _____

1. Circle all clocks that show 11 o'clock.

2. Circle all clocks that show half past 4.

3. Draw or write to show 8 o'clock.

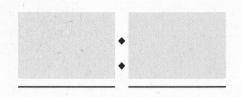

4. Draw or write to show half past 8.

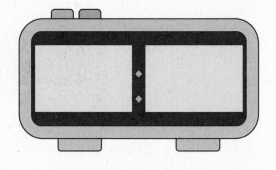

PROBLEM SET

5. Draw lines to match the times.

3 o'clock

1 o'clock

1:00

half past 10

half past 1

2:30

Name

Circle all clocks that show half past 2.

🔍 **16**

Name _____

_____

How many do you think there are? _____

Total _____

Name

Write the missing numbers.

91	101	111
92	102	112
93	103	113
94		
95	105	115
96	106	116
97	107	117
	108	118
99	109	119
100		

Write numbers.

17

Name

1. Write the missing numbers.

81	91	101	113
82	92	102	112
83	93	103	113
84	94	104	114
85	95	105	115
86	96	106	116
87	7	107	117
88	98	108	118
89	99	109	119
90	100	110	120

2. Write the part or the total.

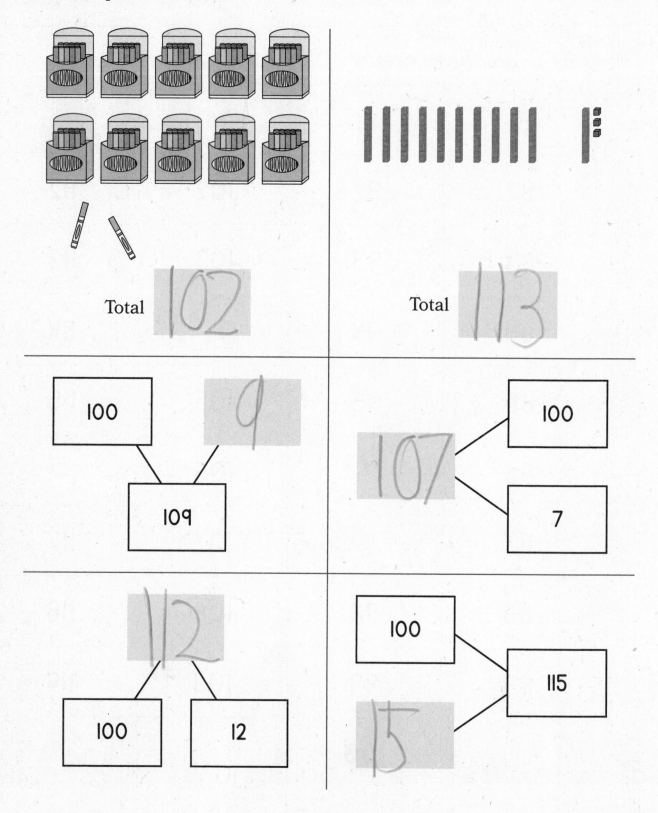

Total 102

Total 113

100 / 109 / 9

107 / 100 / 7

112 / 100 / 12

100 / 115 / 15

PROBLEM SET

3. Draw or write to show 120.

4. Draw or write to show 120 in a new way.

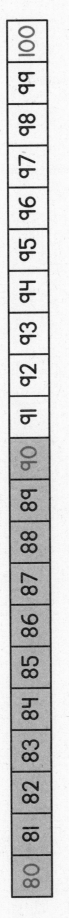

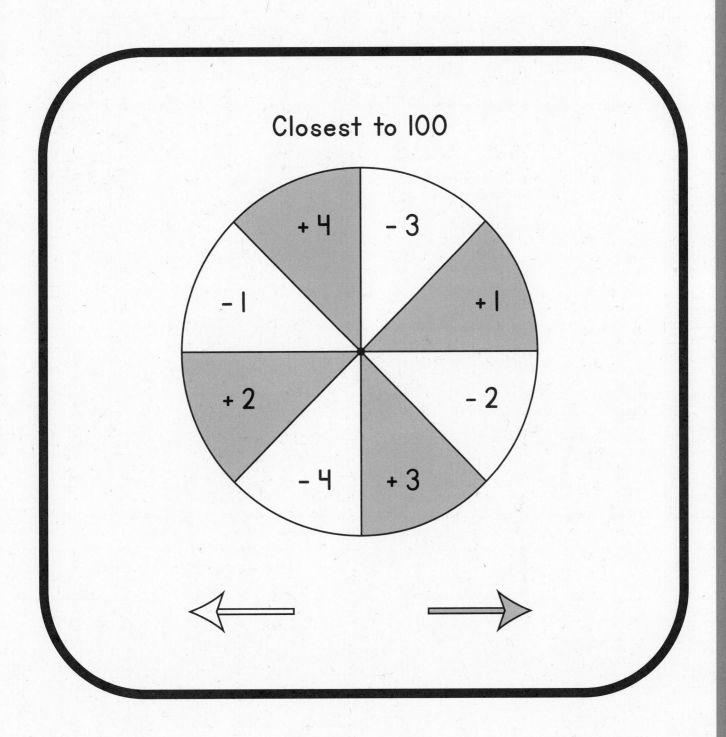

Closest to 100

18

Name _____

99, ____, 101, 102, 103, ____, 105, 106, 107, 108, ____

____, 111, 112, 113, 114, ____, ____, 117, 118, 119, ____

____, 12, 22, ____, 42, 52, ____

____, 70, 80, 90, ____, ____, 120

Name

1. Count up.

Write the numbers.

87, 88, 89, _____, _____, _____, _____, _____

43, 53, 63, _____, _____, _____, _____, _____

107, 108, 109, _____, _____, _____, _____, _____

50, 60, 70, _____, _____, _____, _____, _____

2. Count down.

Write the numbers.

_____, _____, _____, _____, _____, 101, 102, 103

_____, _____, _____, _____, _____, 80, 90, 100

3. Count by ones.

Write the numbers.

_____, _____, 110, _____, _____

_____, _____, 118, _____, _____

4. Count by tens.

Write the numbers.

_____, _____, 90, _____, _____

_____, _____, 95, _____, _____

5. Cross out the number that does not fit.

86, 87, 88, 89, 80	111, 112, 103, 113, 114, 115

6. How close is the number to 100?

Show how you know.

99

80

106

___

☑ **D**

Name _____

1. Fill in the blanks.

97, 98, 99, _____, _____, 102, 103,

104, 105, _____, 107, 108, 109, _____

2. Count.

✎

Write the total.

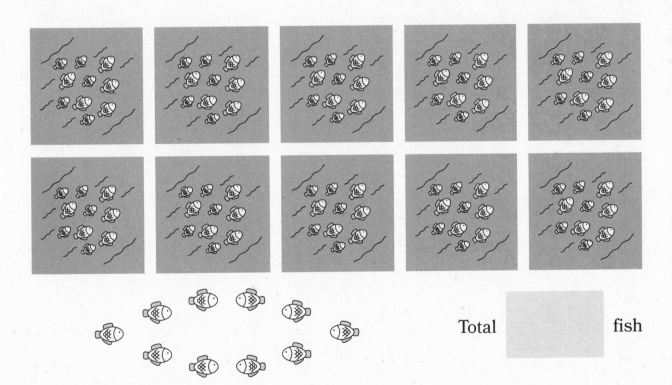

Total [   ] fish

3. Count.

Write the total.

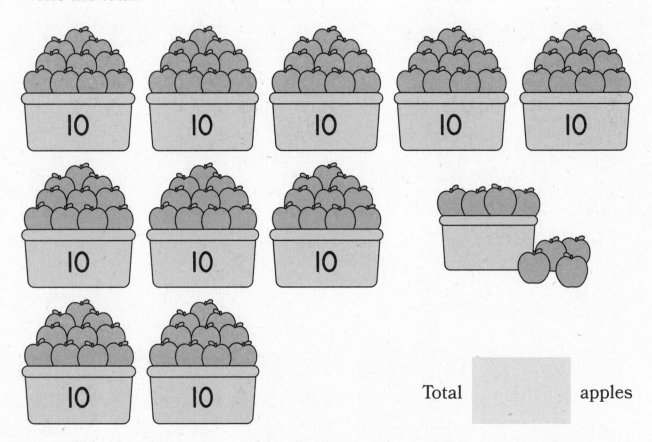

Total ____ apples

4. Draw to show 112.

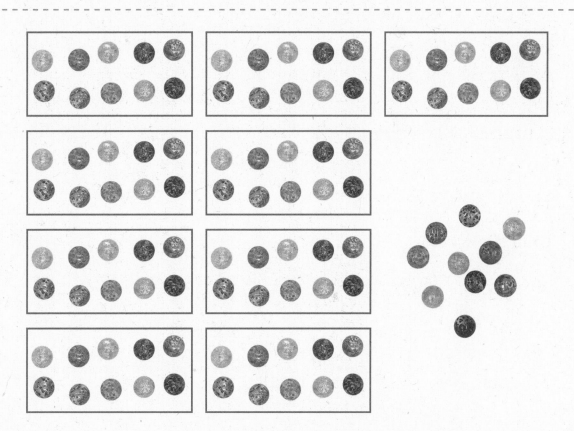

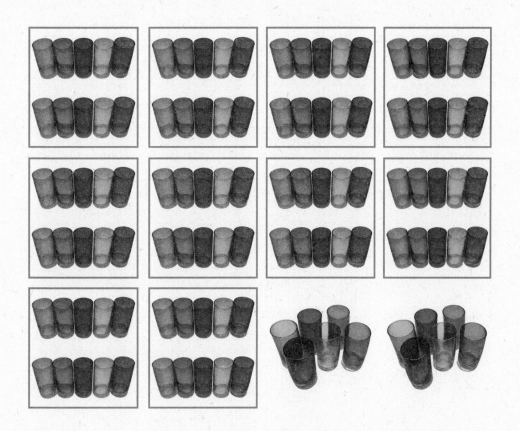

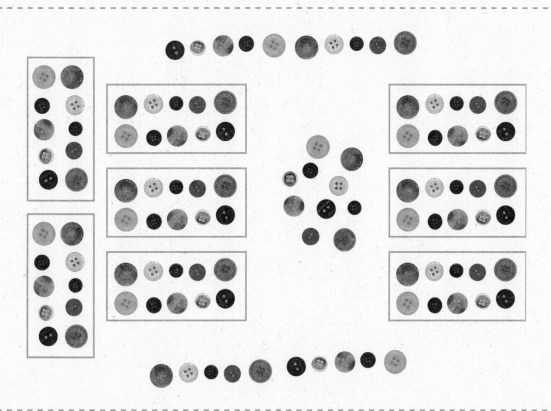

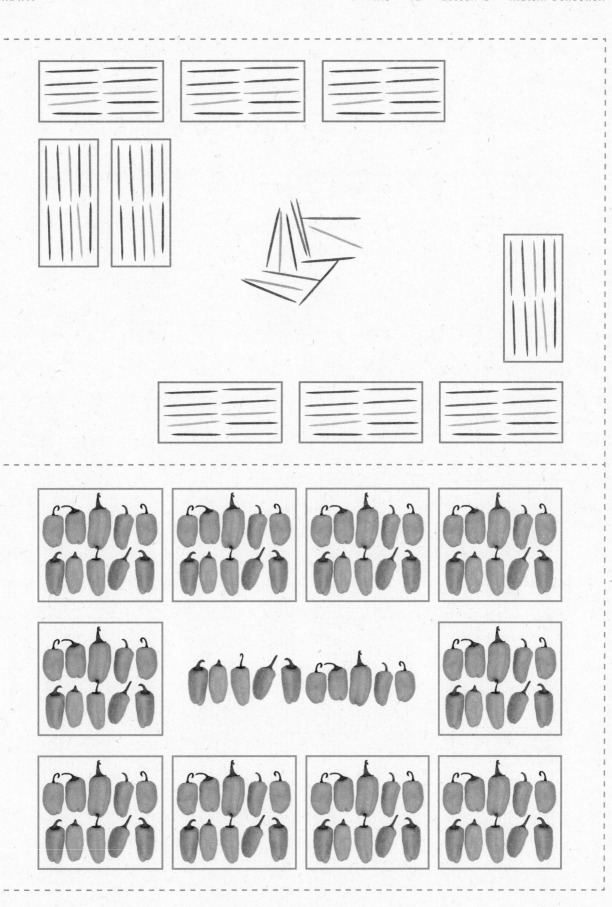

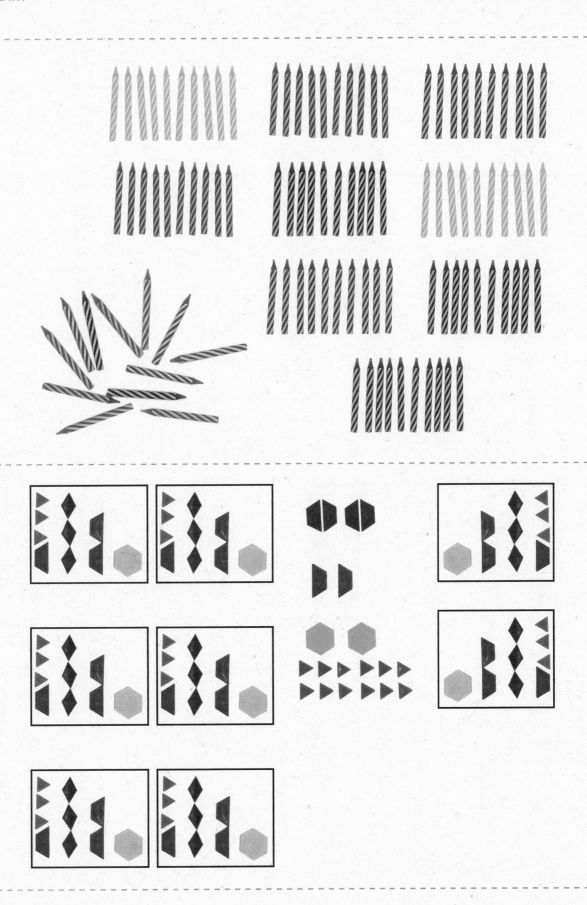

18

Dom

Name

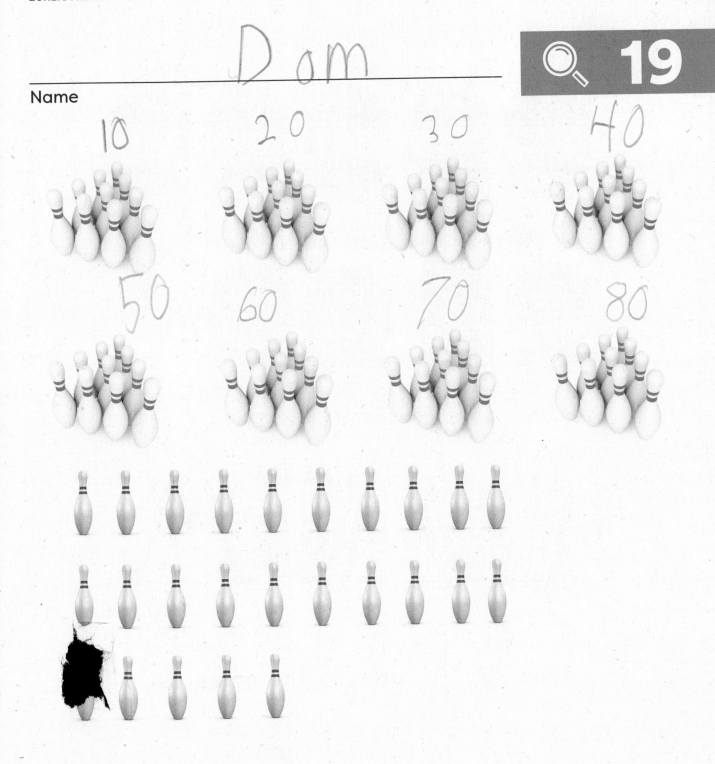

10 20 30 40

50 60 70 80

 8 tens  25 ones is the same as  105 .

 tens  ones is the same as .

3. Write a number greater than 99.

Show it in more than one way.

🔍 **20**

Name _____

Read

Max has 12 tickets.

Kit has 8 tickets.

They need 20 tickets to go on the ride.

Can they ride?

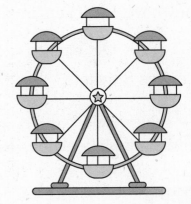

Draw

Write

They have [ ] tickets, so they _____ ride.

Read

Max and Kit are in line for ice cream.

They have 19 tickets in all.

Max has 8 tickets.

How many tickets does Kit have?

Draw

Write

Kit has tickets.

Read

Kit and Dan are in line for the ride.

Kit has 7 tickets.

Dan has 11 tickets.

How many tickets do Kit and Dan have?

Draw

Write

They have [   ] tickets.

Read

Kit and Dan are in line for popcorn.

Dan has 8 tickets.

They have 16 tickets in all.

How many tickets does Kit have?

Draw

Write

Kit has   tickets.

20

Name _____

Read

Mel catches two fish.

How many points does
she have?

Draw

Write

Mel has _____ points.

Zan has 15 points.

He gets the penguin.

How many points does he still have?

Val pops the red balloon.

Then she pops the blue balloon.

How many points does she have?

Name _____

Read

Jade has 12 tickets.

She gets more tickets to go on a ride.

Now she has 20 tickets.

How many tickets did she get?

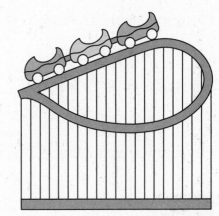

Draw

Write

Jade got _____ tickets.

Read

Nate has 16 tickets.

He uses some tickets to go on a ride.

He has 8 tickets left.

How many tickets did he use?

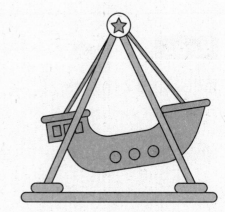

Draw

Write

Nate used tickets.

Name _____

1. **Read**

   Ben gets 9 points.

   Then he gets 3 points.

   How many points does he have?

PRIZES

7   12   15   19

3   6   4

20   9

   Draw

   Write

Ben has _____ points.

 Circle the prizes Ben can choose.

2. **Read**

Baz has 20 points.

He gets the skateboard.

How many points does he
have left?

Draw

Write

Baz has ⬚ point left.

3. **Read**

Tam has 9 points.

She got some more points.

Now she has 15 points.

How many more points did she get?

Draw

Write

Tam got ⬚ more points.

4. **Read**

Wes has 20 points.

He uses some points for a prize.

He has 5 points left.

Which prize did he get?

PRIZES

7   12   15   19

3   6   4

20   9

Draw

Write

Wes used [ ] points to get the   .

5. **Read**

Val wins the skateboard.

How did she get 19 points?

Draw

Write

She hit , , and .

_____

Name

Read

Baz has 16 points.

He gets the top.

How many points does he have left?

Draw

Write

Baz has points left.

Name _____

Read

Baz has some dollars.

He spends 10 dollars.

He still has 2 dollars left.

How many dollars did Baz have to start?

Draw

Write

Baz had   dollars to start.

Read

Kit has some dollars.

She gets 7 more dollars for helping her mom.

Now she has 15 dollars.

How many dollars did Kit have to start?

Draw

Write

Kit had dollars to start.

Name _____

1. **Read**

   Liv got some points.

   Then she got 2 more points.

   Now she has 6 points.

   How many points did she get at first?

   Draw

   Write

   Liv got ⬜ points at first.

2. **Read**

Ren got some points.

Then he got 4 more points.

He got the robot.

How many points did he get at first?

Draw

Write

Ren got ___ points at first.

3. **Read**

Sam has some points.

He uses 6 points to get the ball.

He has 9 points left.

How many points did he have to start?

Draw

Write

Sam had _____ points.

4. **Read**

Ned has some points.

He uses points to get
the glasses.

He has 10 points left.

How many points did he have to start?

Draw

Write

Ned had ⬜ points.

22

_____

Name

Read

Peg got some points.

Then she got 6 more points.

Now she has 11 points.

How many points did she get at first?

Draw

Write

Peg got   points at first.

Name _____

23

1. **Read**

   Tam plays for 10 minutes.

   Max plays for 6 minutes.

   How many more minutes does Tam play than Max plays?

   Draw

   Write

   Tam plays _____ more minutes.

2. **Read**

Math takes 12 minutes.

Art takes 4 minutes.

How many fewer minutes does art take than math takes?

Draw

Activity	Time
	4 minutes
	10 minutes
	12 minutes
	8 minutes
	6 minutes

Write

Art takes ____ fewer minutes.

3. **Read**

Kit plays for 10 minutes.

Ben plays for 4 fewer minutes than Kit plays.

How long does Ben play?

Activity	Time
	4 minutes
	10 minutes
	12 minutes
	8 minutes
	6 minutes

Draw

Write

Ben plays for _____ minutes.

Circle what Ben plays.

4. **Read**

Jon plays ball.

Deb plays for 2 more minutes than Jon plays.

How many minutes does Deb play?

Draw

Activity	Time
	4 minutes
	10 minutes
	12 minutes
	8 minutes
	6 minutes

Write

Deb plays for ___ minutes.

Circle what Deb plays.

23

Name _____

Read

Dan plays for 8 minutes.

Kit plays for 10 minutes.

How many more minutes does Kit play than Dan plays?

Draw

Write

Kit plays for _____ more minutes.

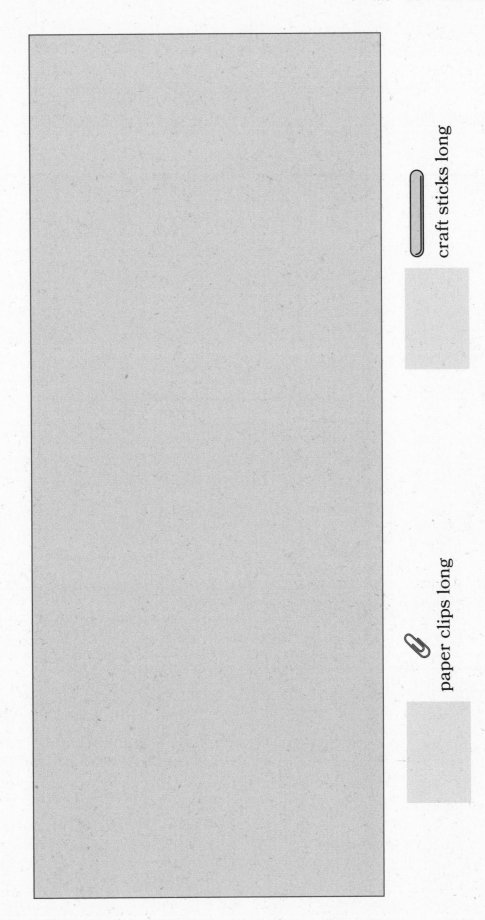

craft sticks long

paper clips long

24

Name

1. Write the total length.

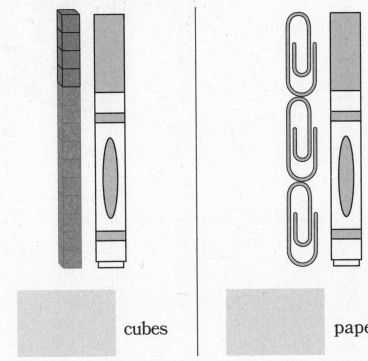

cubes

paper clips

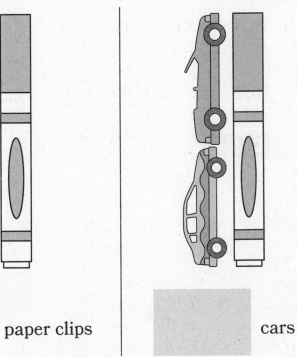

cars

2. Draw or write.

Why do we need **more** cubes than paper clips to measure the marker?

_____

_____

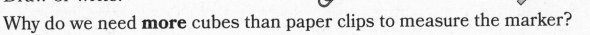

Why do we need **fewer** cars than paper clips to measure the marker?

_____

_____

3. **Read**

Val's hand is 6 paper clips long.

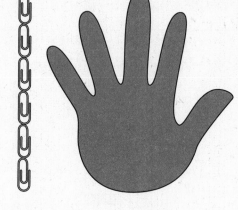

Val's hand is 2 paper clips shorter than her mom's hand.

How long is her mom's hand?

Draw

Write

Her mom's hand is ⬜ paper clips long.

4. **Read**

Val's hand is 6 paper clips long.

Val's hand is 2 paper clips longer than her sister's hand.

How long is her sister's hand?

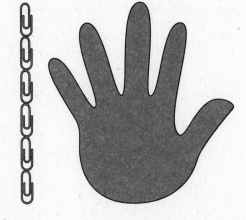

Draw

Write

Her sister's hand is [ ] paper clips long.

Name _____

1. **Read**

   A pencil is 9 paper clips long.

   A book is 16 paper clips long.

   How much longer is the book than the pencil?

   Draw

   Write

   The book is _____  paper clips longer than the pencil.

2. **Read**

Tam has a box of 5 pencils.

He finds some more.

He now has 13 pencils.

How many did he find?

Draw

Write

Tam finds pencils.

Sprint

Subtract.

1.	9 tens – 2 tens	___ tens
2.	90 – 20	___
3.	8 tens – 4 tens	___ tens
4.	80 – 40	___

A

Number Correct: _____

Subtract.

1.	4 tens - 2 tens	___ tens		13.	90 - 40	___
2.	40 - 20	___		14.	70 - 40	___
3.	6 tens - 2 tens	___ tens		15.	70 - 50	___
4.	60 - 20	___		16.	90 - 50	___
5.	8 tens - 2 tens	___ tens		17.	70 - 60	___
6.	80 - 20	___		18.	90 - 60	___
7.	5 tens - 3 tens	___ tens		19.	80 - 70	___
8.	50 - 30	___		20.	90 - 70	___
9.	7 tens - 3 tens	___ tens		21.	90 - 80	___
10.	70 - 30	___		22.	80 - 80	___
11.	9 tens - 3 tens	___ tens		23.	80 - 10	___
12.	90 - 30	___		24.	70 - 30	___

B

Number Correct: _____

Subtract.

1.	3 tens − 2 tens	___ ten
2.	30 − 20	___
3.	5 tens − 2 tens	___ tens
4.	50 − 20	___
5.	7 tens − 2 tens	___ tens
6.	70 − 20	___
7.	4 tens − 3 tens	___ ten
8.	40 − 30	___
9.	6 tens − 3 tens	___ tens
10.	60 − 30	___
11.	8 tens − 3 tens	___ tens
12.	80 − 30	___

13.	80 − 40	___
14.	60 − 40	___
15.	60 − 50	___
16.	80 − 50	___
17.	60 − 60	___
18.	80 − 60	___
19.	80 − 70	___
20.	90 − 70	___
21.	90 − 80	___
22.	80 − 80	___
23.	80 − 10	___
24.	70 − 20	___

Ada Lovelace

Alan Turing

Katherine Johnson

Srinivasa Ramanujan

Mariel Vázquez

Alberto Pedro Calderón

Name

How Many?

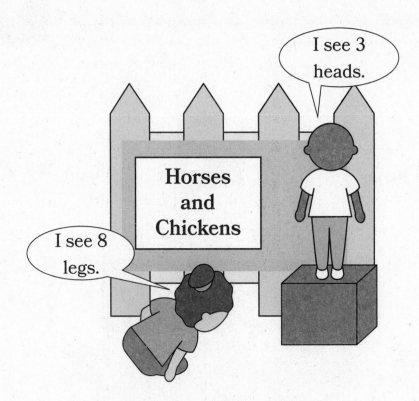

horses   chickens

Name

1. **Read**

How many zebras do they see?

How many flamingos do they see?

Draw

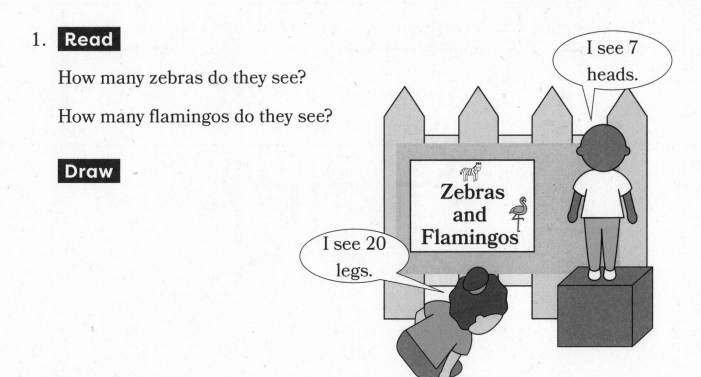

Write

They see [ ] zebras and [ ] flamingos.

2. **Read**

Mel takes care of 6 bats.	Val cares for 3 fewer animals than Mel does. Val's animals cannot fly. 	Jon cares for 10 more animals than Mel does. 

Draw

Write

Name	Animal	How many?
	🦩	
	🦇	
	🐊	

PROBLEM SET

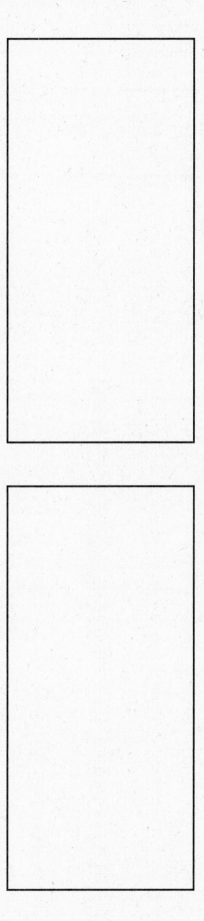

_____ + _____ = 65

_____ + _____ = 65

_____ + _____ = 65

_____ + _____ = 65

_____ + _____ = 65

_____ + _____ = 65

_____ + _____ = 65

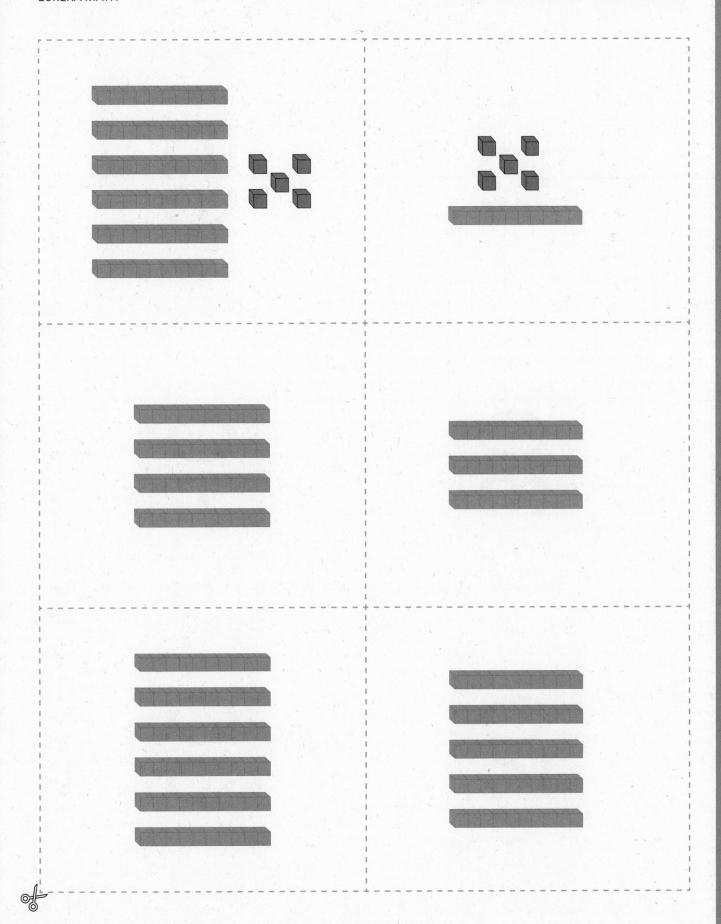

15

65

30

40

50

60

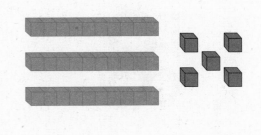

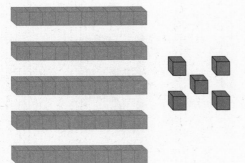

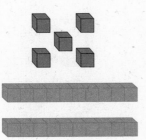

20

0

35

55

5

25

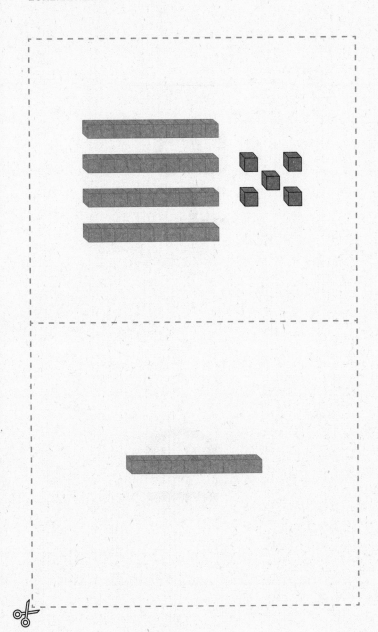

45

10

Name

1. Draw tens and ones to add.

$0 + 55 =$

$10 + 45 =$

$20 + 35 =$

$30 + 25 =$

$40 + 15 =$

$50 + 5 =$

2. Make 75 with two cards.

Show how you know.

70	35	15	50
25	40	5	60

_____ + _____ = 75   _____ + _____ = 75

_____ + _____ = 75   _____ + _____ = 75

PROBLEM SET

3. Write any two-digit number.

Make that number with two parts in different ways.

Show how you know.

☑ 26

Name

Draw tens and ones to add.

$10 + 35 =$

$20 + 25 =$

$30 + 15 =$

Name

1. Add.

Show how you know.

13 + 17 = [ ]

63 + 17 = [ ]

63 + 27 = [ ]

2. Add.

Show how you know.

65 + 33 =

_____

79 + 15 =

3. Write a problem with two-digit numbers.

Show two ways to add.

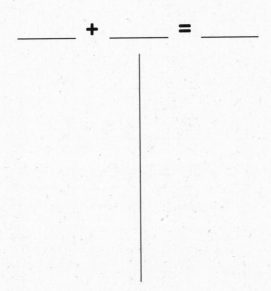

$$\rule{3cm}{0.4pt} + \rule{3cm}{0.4pt} = \rule{3cm}{0.4pt}$$

27

Name _____

Add.

Show how you know.

29 + 32 =

31 + 19 =

57 + 37 =

29 + 55 =

67 + 31 =

29 + 35 =

12 + 66 =

24 + 36 =

46 + 28 =

72 + 19 =

45 + 16 =

11 + 59 =

27 + 29 =

52 + 29 =

38 + 39 =

21 + 48 =

Name

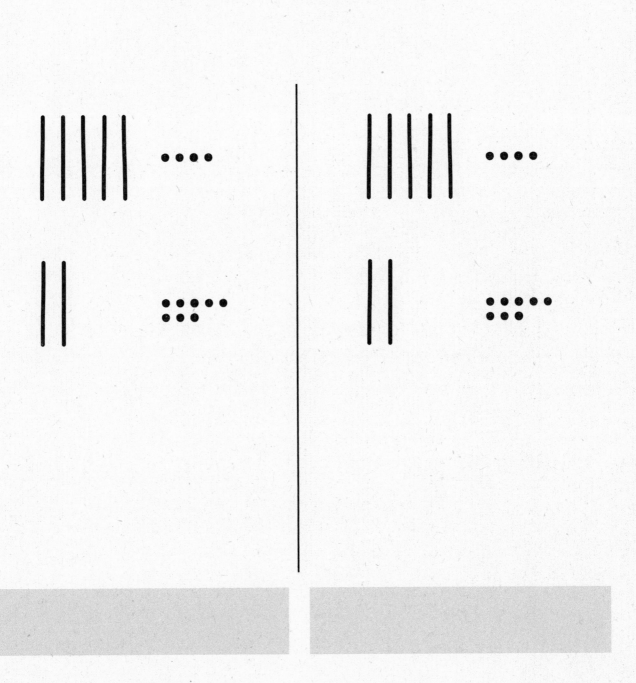

28

Name

1. Add two ways.

$$66 + 33 = \boxed{}$$

$$44 + 36 = \boxed{}$$

2. Add two ways.

$$48 + 25 = \underline{}$$

$$53 + 39 = \underline{}$$

3. Write a problem with 2 two-digit numbers.

Show two ways to add.

_____ + _____ = _____

28

Name _____

Add.

Show how you know.

59 + 33 =

_____ ten + _____ tens = _____ tens

_____ tens + _____ tens = _____ tens

_____ tens + _____ tens = _____ tens

_____ tens + _____ tens = _____ tens

_____ tens + _____ tens = _____ tens

Name _____

Read

Mr. West has some boxes of 10 pencils.

He buys 6 more boxes of 10 pencils.

Now he has 100 pencils total.

How many pencils does he have at first?

Draw

Write

Mr. West has _____ pencils at first.

Name

Read

Val has 4 dimes.

She finds 6 more dimes.

How much money does Val have?

Draw

Read

Ren has 10 bags of apples.

There are 10 apples in each bag.

Ren sells 2 bags of apples.

How many apples does Ren still have?

Draw

Write

Val has _____ cents.

Write

Ren has _____ apples.

Read

Nate puts some books on the shelf.

He puts 10 books in his desk.

There are 100 books total.

How many books did Nate put on the shelf?

Draw

Write

Nate put [ ] books on the shelf.

Read

There are 100 flowers in packs of 10.

Some packs are yellow.

Some packs are blue.

[ ] packs are yellow.

[ ] packs are blue.

Draw

Write

There are [ ] flowers.

_____

Name

Add.

Show how you know.

50 + 50 = ◻

60 + ◻ = 100

1	2	3	4	5	6	7	8	9	10	11	12	13	14	15	16	17	18	19	20

21	22	23	24	25	26	27	28	29	30	31	32	33	34	35	36	37	38	39	40

41	42	43	44	45	46	47	48	49	50	51	52	53	54	55	56	57	58	59	60

61	62	63	64	65	66	67	68	69	70	71	72	73	74	75	76	77	78	79	80

81	82	83	84	85	86	87	88	89	90	91	92	93	94	95	96	97	98	99	100

＝ ＋	＝ ＋
＝ ＋	＝ ＋

Name

1. Add.

Use your number path.

$54 + 6 + 40 =$

$43 + 7 + 50 =$

$14 + 6 + 80 =$

$23 + 7 + 70 =$

2. Make 100 with your number path.

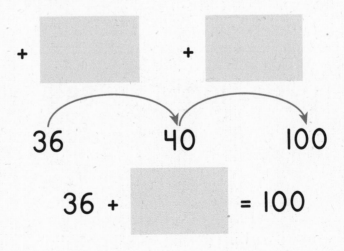

$+$ $+$

36   40   100

$36 +$ $= 100$

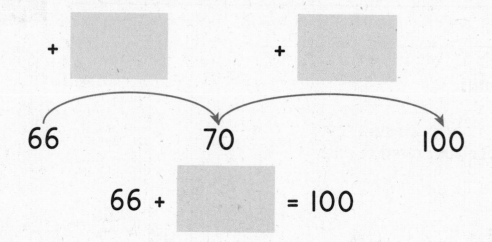

$$66 + \boxed{} = 100$$

3. Find the unknown part.

Show how you know.

$$54 + \boxed{} = 100 \qquad 68 + \boxed{} = 100$$

$$77 + \boxed{} = 100 \qquad 89 + \boxed{} = 100$$

_____

Name

Find the unknown number.

Show how you know.

68 + ▢ = 100

Sprint

Add.

1.	3 tens + 2 tens	tens
2.	30 + 20	
3.	20 + 5	
4.	20 + 15	

A

Number Correct: _____

Add.

1.	4 tens + 1 ten	tens
2.	40 + 10	
3.	5 tens + 2 tens	tens
4.	50 + 20	
5.	6 tens + 3 tens	tens
6.	60 + 30	
7.	30 + 60	
8.	20 + 50	
9.	10 + 40	
10.	20 + 40	
11.	30 + 50	
12.	40 + 60	

13.	40 + 5	
14.	40 + 15	
15.	50 + 15	
16.	50 + 25	
17.	60 + 25	
18.	60 + 35	
19.	35 + 60	
20.	25 + 60	
21.	15 + 70	
22.	25 + 70	
23.	35 + 50	
24.	45 + 50	

B

Number Correct: _____

Add.

1.	3 tens + 1 ten	tens
2.	30 + 10	
3.	4 tens + 2 tens	tens
4.	40 + 20	
5.	5 tens + 3 tens	tens
6.	50 + 30	
7.	30 + 50	
8.	20 + 40	
9.	10 + 30	
10.	20 + 30	
11.	30 + 40	
12.	40 + 50	

13.	30 + 5	
14.	30 + 15	
15.	40 + 15	
16.	40 + 25	
17.	50 + 25	
18.	50 + 35	
19.	35 + 50	
20.	25 + 50	
21.	15 + 60	
22.	25 + 60	
23.	35 + 40	
24.	45 + 40	

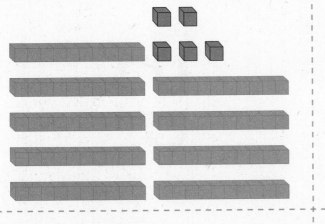

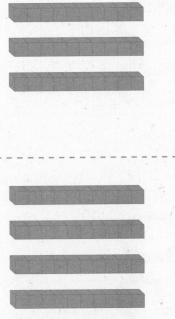

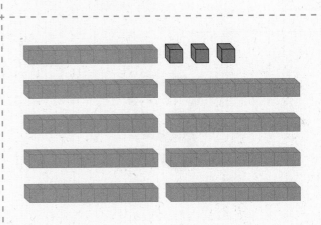

1

7

30

95

50

70

93

99

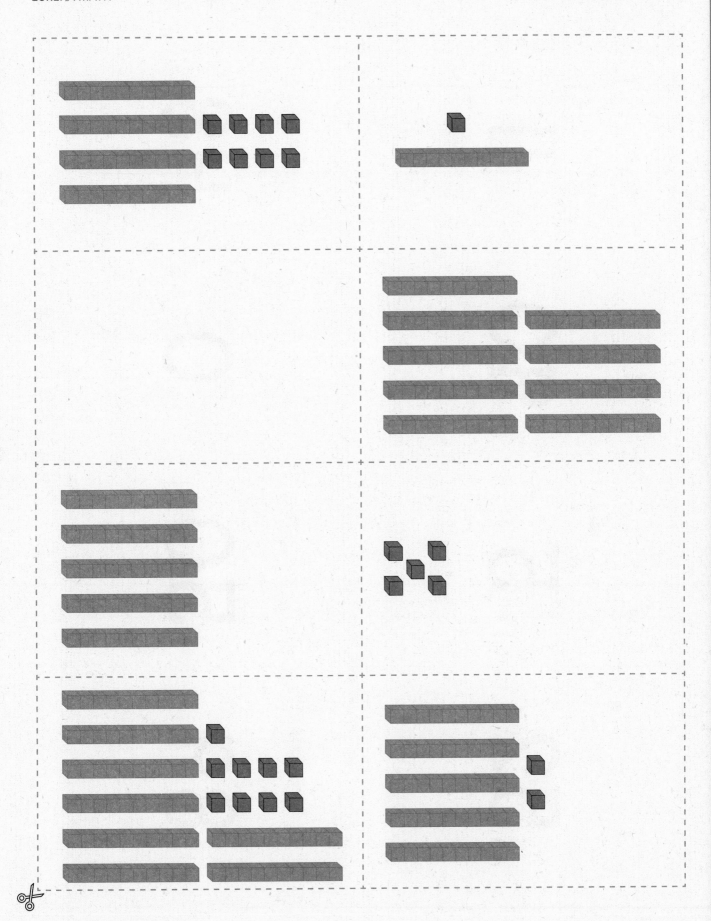

11

48

90

0

5

50

52

89

10

100

_____ + _____ = 100

_____ + _____ = 100

_____ + _____ = 100

_____ + _____ = 100

_____ + _____ = 100

_____ + _____ = 100

_____ + _____ = 100

_____ + _____ = 100

_____ + _____ = 100

Make your own match.

_____ + _____ = 100

Name

F

Name _____

1. Add.

36 + 64 = ▢   44 + 55 = ▢

2. Draw or write.

What did you learn in math this year?	What are you good at in math?
What is hard for you in math?	What do you like about math?

Credits

Great Minds® has made every effort to obtain permission for the reprinting of all copyrighted material. If any owner of copyrighted material is not acknowledged herein, please contact Great Minds for proper acknowledgment in all future editions and reprints of this module.

Cover, Edward Hopper (1882–1967), *Tables for Ladies*, 1930. Oil on canvas, H. 48-1/4, W. 60-1/4 in. (122.6 x 153 cm.). George A. Hearn Fund, 1931 (31.62). The Metropolitan Museum of Art. © 2020 Heirs of Josephine N. Hopper/Licensed by Artists Rights Society (ARS), NY. Photo Credit: Image copyright © The Metropolitan Museum of Art. Image source: Art Resource, NY; page 45, robuart/ Shutterstock.com, design56/Shutterstock.com, tomeqs/Shutterstock, BlueRingMedia/Shutterstock. com; page 111 (from top left), AlexeiLogvinovich/Shutterstock.com, asiandelight/Shutterstock. com, bergamont/Shutterstock.com, kosam/Shutterstock.com; pages 125, 130, Krakenimages.com/ Shutterstock.com; page 185, (composite image) doomu/Shutterstock.com, NiglayNik/Shutterstock. com; page 237, (composite image) Margaret Sarah Carpenter, Portrait of Ada Lovelace, 1836. and Photograph of Srinivasa Ramanujan. Public domain via Wikimedia Commons, famouspeople/Alamy Stock Photo, Photo credit: University of California Davis. Photo by Gregory Urquiaga, Photograph Courtesy of the University of Chicago, Courtesy NASA/Bob Nye, Intellson/Shutterstock.com; page 242, (top) BlueRingMedia/Shutterstock.com, (bottom), subarashii21/Shutterstock.com; All other images are the property of Great Minds.

For a complete list of credits, visit http://eurmath.link/media-credits.

Acknowledgments

Kelly Alsup, Dawn Burns, Jasmine Calin, Mary Christensen-Cooper, Cheri DeBusk, Stephanie DeGiulio, Jill Diniz, Brittany duPont, Melissa Elias, Lacy Endo-Peery, Scott Farrar, Krysta Gibbs, Melanie Gutierrez, Eddie Hampton, Tiffany Hill, Robert Hollister, Christine Hopkinson, Rachel Hylton, Travis Jones, Kelly Kagamas Tomkies, Liz Krisher, Ben McCarty, Maureen McNamara Jones, Cristina Metcalf, Ashley Meyer, Melissa Mink, Richard Monke, Bruce Myers, Marya Myers, Andrea Neophytou Hart, Kelley Padilla, Kim L. Pettig, Marlene Pineda, Elizabeth Re, John Reynolds, Meri Robie-Craven, Robyn Sorenson, Marianne Strayton, James Tanton, Julia Tessler, Philippa Walker, Lisa Watts Lawton, MaryJo Wieland

Trevor Barnes, Brianna Bemel, Adam Cardais, Christina Cooper, Natasha Curtis, Jessica Dahl, Brandon Dawley, Delsena Draper, Sandy Engelman, Tamara Estrada, Soudea Forbes, Jen Forbus, Reba Frederics, Liz Gabbard, Diana Ghazzawi, Lisa Giddens-White, Laurie Gonsoulin, Nathan Hall, Cassie Hart, Marcela Hernandez, Rachel Hirsh, Abbi Hoerst, Libby Howard, Amy Kanjuka, Ashley Kelley, Lisa King, Sarah Kopec, Drew Krepp, Crystal Love, Maya Márquez, Siena Mazero, Cindy Medici, Ivonne Mercado, Sandra Mercado, Brian Methe, Patricia Mickelberry, Mary-Lise Nazaire, Corinne Newbegin, Max Oosterbaan, Tamara Otto, Christine Palmtag, Andy Peterson, Lizette Porras, Karen Rollhauser, Neela Roy, Gina Schenck, Amy Schoon, Aaron Shields, Leigh Sterten, Mary Sudul, Lisa Sweeney, Samuel Weyand, Dave White, Charmaine Whitman, Nicole Williams, Glenda Wisenburn-Burke, Howard Yaffe

Talking Tool

I Can Share My Thinking	My drawing shows . . . . I did it this way because . . . . I think ____ because . . . .
I Can Agree or Disagree	I agree because . . . . I disagree because . . . . I did it a different way. I . . . .
I Can Ask Questions	How did you . . . ? Why did you . . . ? Can you explain . . . ?
I Can Say It Again	I heard you say . . . . ____ said . . . . Can you say it another way?

Thinking Tool

When I work on a task, I ask myself

Before	Have I done this before? What strategy will I use? Do I need any tools?
During	Is my strategy working? Should I try something else?
After	What worked well? What did not work?

At the end of each class, I ask myself

	What did I learn? Do I have any questions?